M1 ABV

Assault Breacher Vehicle

Warmachines 01 is a visual reference of the U.S. Army and U.S. Marine Corps M1 Abrams-based assault breacher vehicle.

VP
ERLINDEN PUBLICATIONS

SABOT Publications

Warmachines 01 M1 ABV

SABOT Publications
941 Cottontail Lane
St. Charles, Missouri 63303
United States of America
ISBN: 978-0-9973774-6-0

Special thanks to our friends for their valuable contributions to this book:
Francois Verlinden - for his inspiration and dedication to the hobby for many years
CSM Mike Moske
SPC Cedric Ellington Squire
Robert Burik
SFC Timothy Gwynn
SSG Shiran Jackson
John Charvat
Mike Petty
SPC Shannon Khoury
Combat engineers, Tankers and Mechanics everywhere
and all of the Soldiers, Sailors, Marines and Airmen who serve
Thank you!

Chris Mrosko
Brett Avants
Pearce Browning
SABOT Publications LLC

Printed in the United States of America

Introduction

The M1 Assault Breacher Vehicle (ABV) is a U.S. military combat engineering and obstacle clearing vehicle based on the M1 Abrams chassis. It is currently used by both the U.S. Army and the United States Marine Corps (USMC). It was especially designed to clear lanes for troops and vehicular traffic through minefields, obstacles and roadside bombs and improvised explosive devices (IEDs). It carries a crew of 2.

The ABV It consists of an M1 Abrams tank hull, a unique turret, a Lane Marking System (LMS), Integrated Vision System, and a High Lift Adapter that interchangeably mounts a Full Width Mine Plow (FWMP) or a Combat Dozer Blade (CDB). The M1 ABV is also armed with a Mine Clearing Line Charge (MCLC) system which uses an explosive-tipped rocket that launches a long string of C4 explosive charges for clearing long lanes through minefileds and other obstacles.

The first use of the ABV in combat was during the USMC *Operation Cobra's Anger* in the Helmand province southwest of Kabul, Afghanistan, in December 2009. In February 2010, the USMC also used M1 ABV's as part of *Operation Moshtarek* against the Taliban around Marjah, Afghanistan. M1 ABV's have also been deployed to South Korea.

The ABV improves the mobility and survivability of combat engineers while having the speed and ability to keep pace with the maneuver force. It creates a tank-width cleared lane through a minefield by launching and detonating one of its MCLC systems across the minefield, then proofing the lane with its FWMP while marking the cleared lane with its LMS. The ABV is air-transportable by C-17 and larger aircraft.

The ABV with the plow is know as THE SHREDDER, named after a Tennage Mutant Ninja Turtle character! The ABV with the dozer blade is known as BLADE.

ABOVE: ABV operators with 2nd Combat Engineer Battalion, 2nd Marine Division, stage their ABVs during a mechanized breach drill. The smoke grenade launchers indicate that this is a USMC ABV. BELOW: A U.S. Marine Corps ABV assigned to Mobility Assault Company, 1st Combat Engineer Battalion, clears heavy debris from a path for an awaiting convoy while conducting complex obstacle breaching exercises during Steel Knight 2014 (SK-14) on Camp Pendleton, CA, Dec. 10, 2013.

BELOW LEFT: Soldiers must load lane marking poles into the Lane Marking System (LMS) equipped on either side of the ABV to simultaneously work with the FWMP during operations. BELOW RIGHT: The USMC tank commander gives instruction to his driver down below.

ABOVE and BELOW: Marines with 2nd Combat Engineer Battalion, 2nd Marine Division use breaching equipment to get past one of the obstacles during the Breaching and clearing exercise Nov. 13, 2013 at Marine Corps Base Camp Lejeune.

BELOW: A U.S. Marine Corps ABV assigned to Mobility Assault Company, 1st Combat Engineer Battalion, proceeds over barriers via a path provided by a M60A1 Armored Vehicle-Launched Bridge while conducting complex obstacle breaching exercises during Steel Knight 2014 (SK-14) on Camp Pendleton, CA, Dec. 10, 2013. The High Lift Adapter (HLA) on the front end allow for the ABV to cross the bridge without damaging the Front-End Equipment (FEE).

ABOVE: U.S. Soldiers with Charlie Company, 1st Brigade Special Troops Battalion, 1st Armored Brigade Combat Team, 2nd Infantry Division dig through the ground searching for mines in a simulated minefield with the ABV during a capabilities demonstration at Local Training Area Watkins in Dongducheon, South Korea, Aug. 18, 2013. BELOW: A U.S. Marine ABV assigned to 2nd Combat Engineer Battalion (CEB) based out of Camp Lejeune, participates in 1st Battalion, 4th Marine Regiment's Battalion Assault Course (BAC) during Integrated Training Exercise (ITX) 2-15 at Blacktop Training Area at Camp Wilson, Marine Corps Air Ground Combat Center Twentynine Palms, CA, Feb. 9th, 2015.

ABOVE: Engineers with Company C, 1st Brigade Special Troops Battalion, 1st Armored Brigade Combat Team, 2nd Infantry Division, prepare to fire the mine clearing line initiating charge in the ABV during a demonstration held at Rodriguez Live Fire Range. BELOW: Soldiers with Company C, 1st Brigade Special Troops Battalion, 1st Armored Brigade Combat Team, 2nd Infantry Division, clear a simulated minefield with the ABV during a capabilities demonstration at Local Training Area Watkins, South Korea.

ABOVE: U.S. Soldiers of Company A, 10th Brigade Engineer Battalion, 1st Armored Brigade Combat Team move their ABV behind cover while conducting troop movement training during exercise Combined Resolve IV at the U.S. Army's Joint Multinational Readiness Center in Hohenfels, Germany, May 27, 2015. BELOW: Marines with Mobile Assault Company, 2nd Combat Engineer Battalion, plow dirt in an ABV during a vehicle breaching exercise aboard Engineer Training Area 2, Camp Lejeune, N.C.

ABOVE: U.S. Army AVB on exercise. Note the improvised camouflage of bare branches and some pine branches. Also notice the green drive sprocket teeth and the two green side skirts. BELOW: MCLC is set to the proper angle as engineers approach the objective. Again, note the sparse camouflage from branches and the green side skirts.

ABOVE and RIGHT: If it fits, then it sits. In this case, it's one-hundred meters of C4 Training Blocks sitting in each linear charge crate.

The MCLC in mid-flight as it fully extends the explosives content across the minefield.

Soldiers are tasked with inspecting and re-packing the linear charge into the crate before putting it to work.

ABOVE: The ABV on training exercises at Ft. Leonard Wood, Missouri, in April 2016. The mud and red clay make for a tough environment for the ABV's blade. BELOW: The ABV strikes a pose alongside the Armored Combat Earthmover (ACE) at Ft. Leonard Wood.

ABOVE: Operators train on approaching, crossing, and dismounting techniques with the M60 Armored Vehicle-Launched Bridge (AVLB). BELOW: The heavy terrain of mud and red clay at Ft. Leonard Wood proves to be no match for the ABV. Note the build-up on the wheels and sprocket.

NO STEP
AABV009

NO STEP
B512

Pages 14 - 15: Another look at the ABV with one of its two MCLCs raised to the safe firing angle prior to launch. The blue color indicates an inert training round.

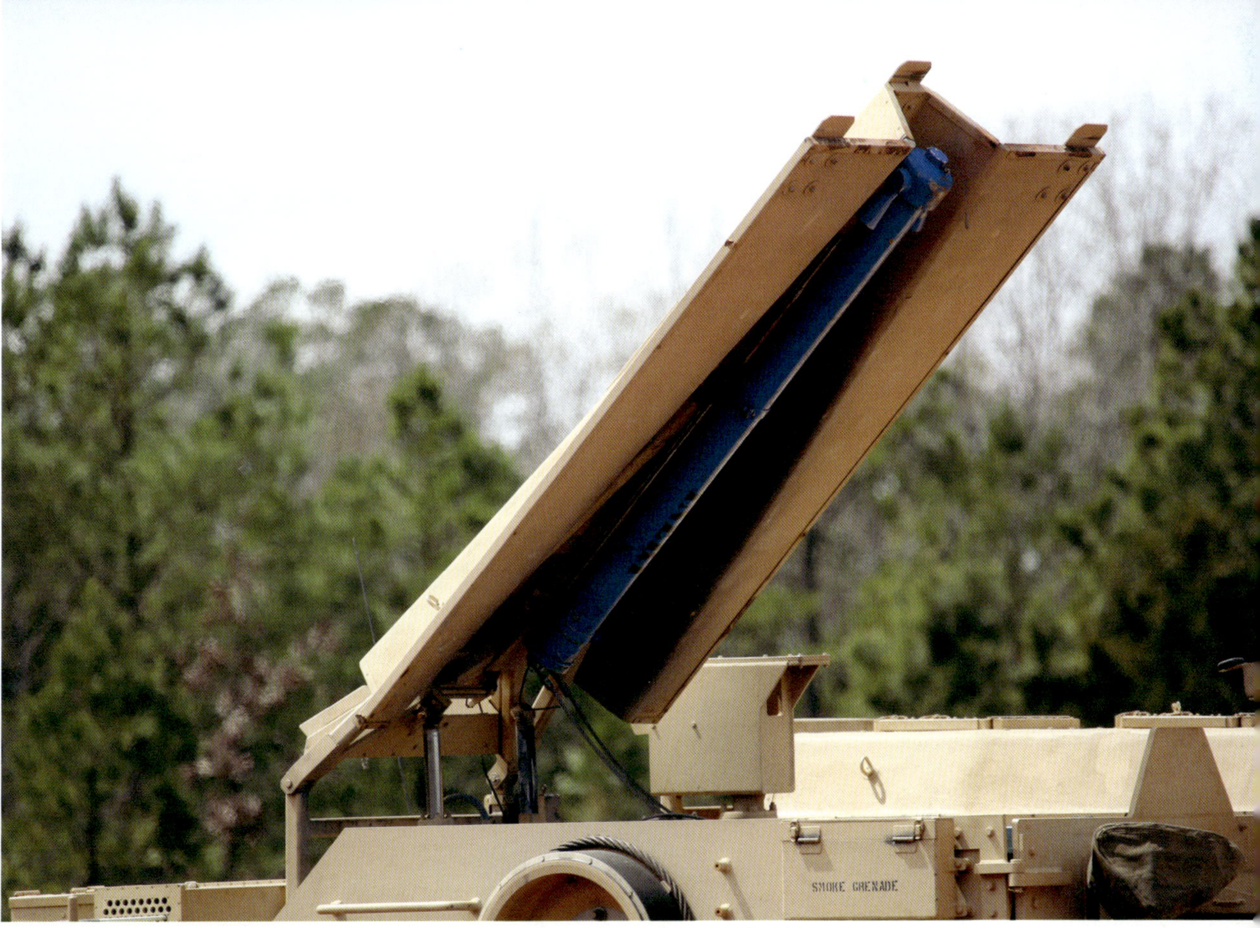

ABOVE and BELOW: The ABV with it's FWMP fully submerged in the dirt and ground of a training facility. This vehicle is clearing a path for a follow-on convoy. Note the extended LMS on the side, ready to mark the lane boundaries for subsequent vehicles.

ABOVE: Close-ups of the mud and clay at Ft. Leonard Wood. Note the holes in the sprocket that allows the mud to work its way out, minimizing the possibility of thrown track due to clogging. BELOW: A close view of the FWMP.

The ABV operators spend hours in the field training on proper lane clearing procedures using the FWMP along with 'dummy mines' for accurate evaluations of the drivers handling.

ABOVE: Trainees at Fort Leonard Wood watch as their fellow soldiers conduct training with the Combat Dozer Blade. Note the weathering on the blade, the track and the side skirts from the clay. BELOW: The tank commander guide's his driver as they make their way through a course full of barriers and other obstacles. This vehicle just pushed this barrier out of the way with the blade attachment.

ABOVE and BELOW: Take a closer look at the air-pressured LMS, each one capable of carrying up to fifty lane-marking poles. These are used to mark the boundaries of the lane cleared by the ABV for subsequent vehicular and foot traffic.

A right side shot of the rear half of the ABV, showing the deployed and empty LMS, along with the spare roadwheel and a tow cable wrapped around it. Even thhough this vehicle has been through training in teh mud and red clay, not the cleanliness of the side skirts.

ABOVE: With the Combat Dozer Blade attached, the ABV is capable of executing similar tasks as that of the M9 ACE. BELOW: A close-up view of the blade attachment. This one is still clean!

ABOVE and BELOW: Soldiers put the ABV to the test as they conduct grading operations utilizing the Combat Dozer Blade at Ft. Leonard Wood, MO, in April 2016. The blade can be used to excavate dirt and ground for lane-clearing, or to move obstacles and barriers out of the way for follow-on vehicles and troops.

ABOVE: After continuous rounds of carefully applied technique with the dozer, engineers were able to successfully level the surface for easy access by future convoys. BELOW: With so much force behind its engine, the ABV can easily push terrain to create defilade, berms, and other various type of defensive structures as seen here.

ABOVE and BELOW: The 'teeth' on the FWMP allow mines to surface so the outward-angled blades can push them away from the lane with a low chance of detonation. Note the dirt and red clay in the teeth, on the skids and on the blade itself of this mine plow.

ABOVE: Another oblique view of the FWMP. BELOW: The rear of the ABV. The track pads on this vehicle are worn down to the metal track.

Pages 28 - 29: Close up views of the various aspects of the FWMP. Note how the dirt and ground accumulates on the various parts of the mine plow skids, the plow and in the teeth.

Pages 30 - 31: More close-ups of the FWMP. The photos on the bottom of pages 30 and 31 show the plow detached from the ABV and sitting in the motor pool of Ft. Leonard Wood. Note the attachment points and systems on the plow.

ABOVE and BELOW: Two views of the detached dozer blade. The jacks are able to be stowed on the Combat Dozer Blade during travel which allows operators to detach the equipment at any place or time.

ABOVE and BELOW: Two close-up views of the blade. When attaching the Front-End Equipment to the ABV these cables connect to the vehicle's hydraulic system which allows the operator to maneuver the equipment while locked in the High-Lift Adapter (HLA).

ABOVE and BELOW: With the HLA locked into place, the blade is fully raised and prepared for travel. Note in the photo above how the blade is attached to the front of the ABV.

ABOVE: The jacks are tucked away right behind the blade and secured with quick release pins. BELOW: A nice full right-side view of the ABV showing the blade attached to the front of the vehicle.

ABOVE: Looking down the hatch into the tank commander's position. BELOW: The driver in the driver's compartment. Note the video screen in front, which allows the driver to get a better view of the battlefield while buttoned up.

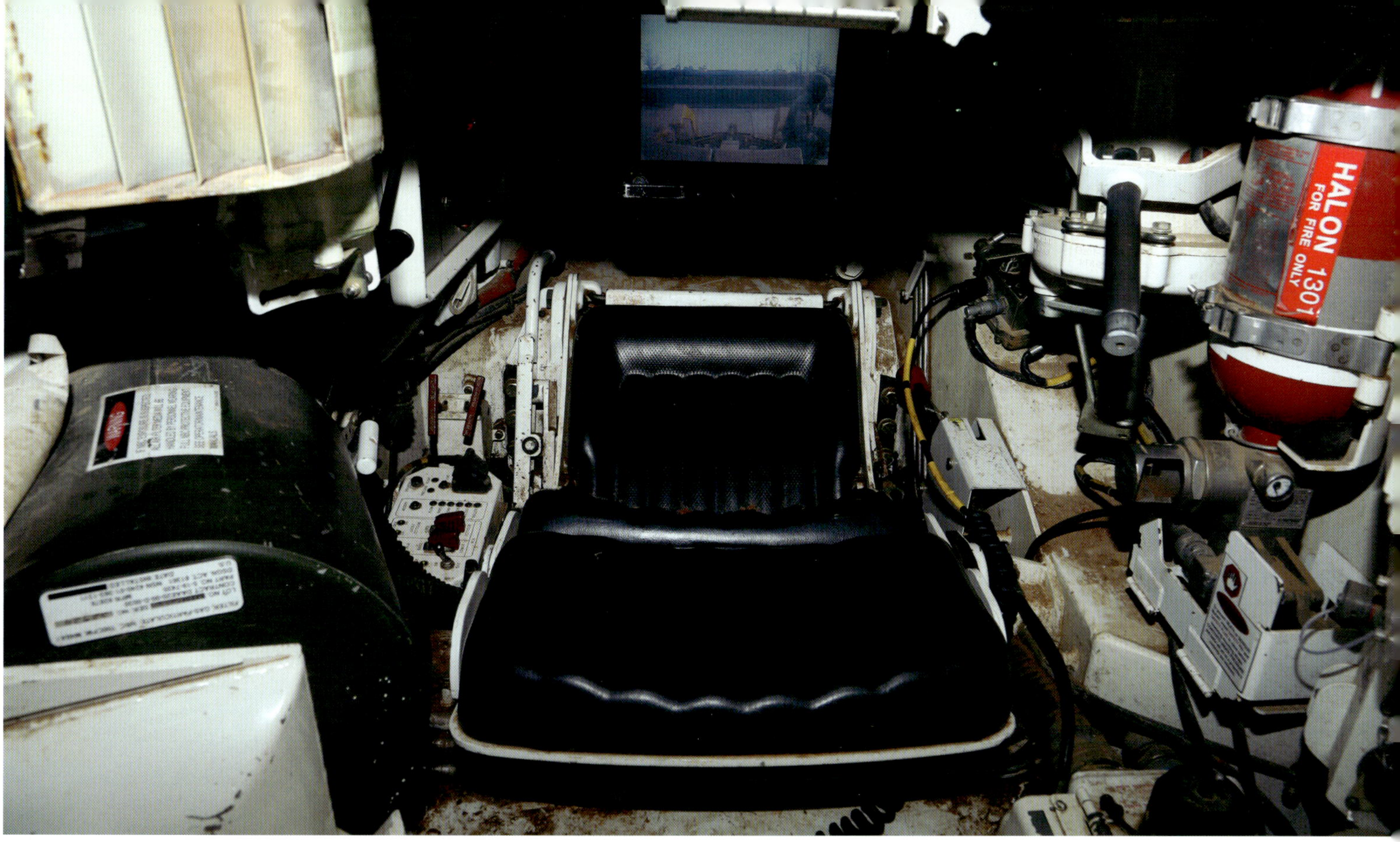

ABOVE: The driver's compartment. You can see the front of the vehicle on the video screen. The driver's area is pretty cramped. BELOW: Another view of the tank commander's position.

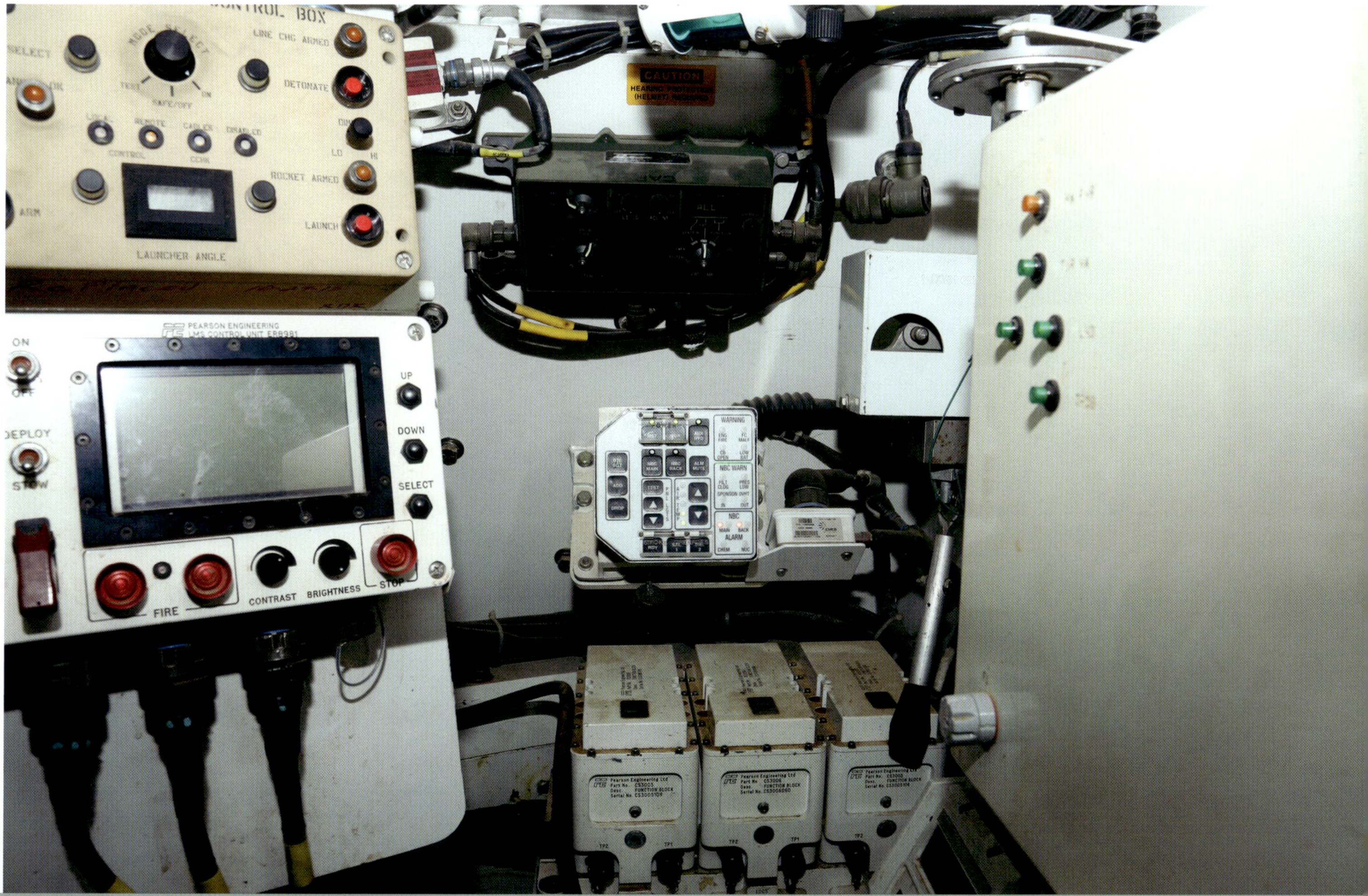

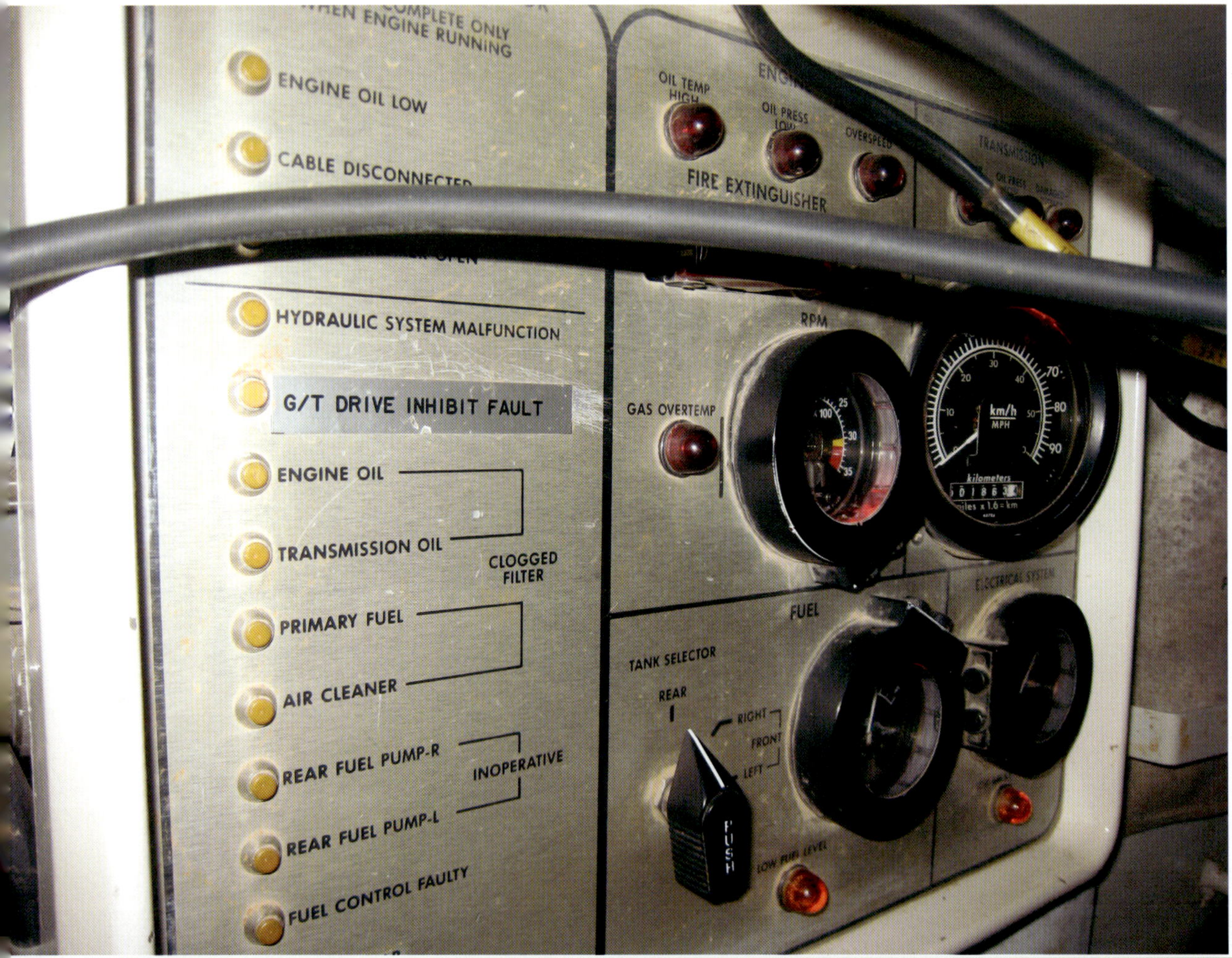

ABOVE: The control panels for the MCLC and the LMS are seen here on the top and bottom, respectively. LEFT: The driver's instrument panel (DIP) allows the driver to monitor the functionality of the vehicle and pin-point a problem prior to performing troubleshooting procedures.

ABOVE: From the driver's master panel (DMP), operators can control simple accessories such as the lights, start or stop the engine, and power up the integrated vision system (IVS). RIGHT: Like most military vehicles, the ABV uses an amplifier to regulate power for communication devices to provide high-quality external communication. It is also equipped with an internal communication system that allows the driver and tank commander to hear each other.

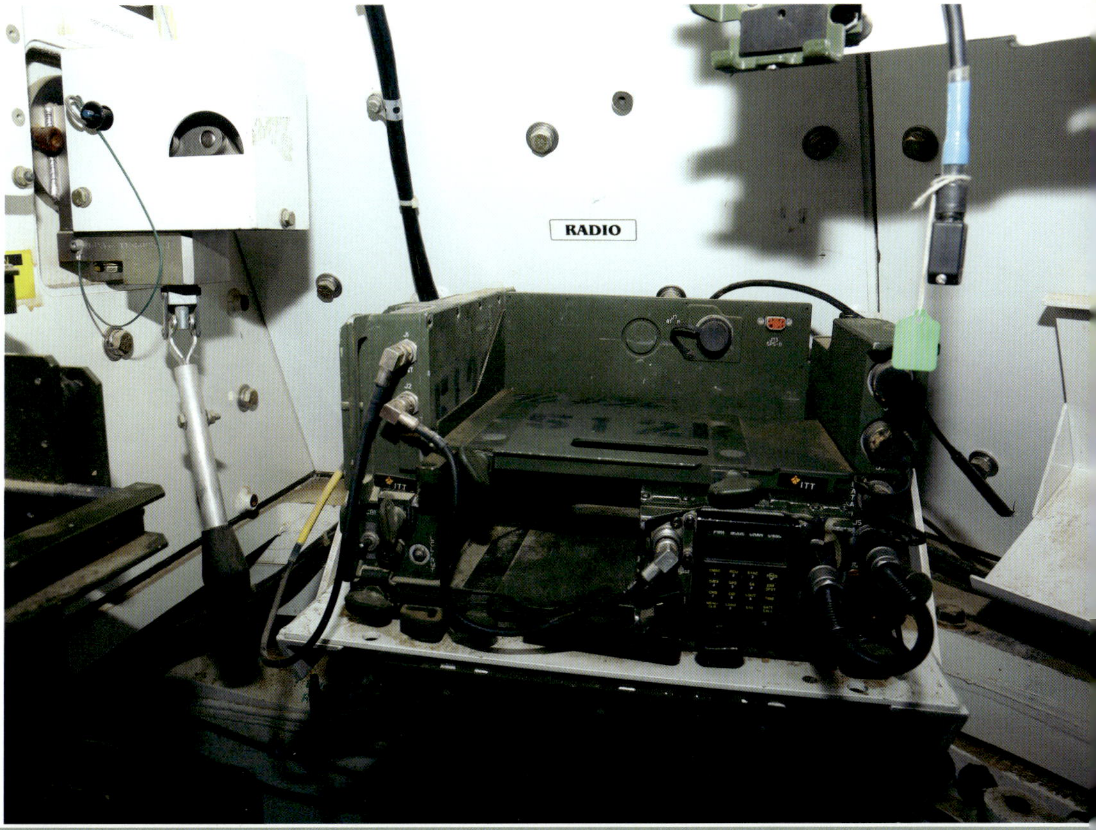

ABOVE: Top view of the two MCLC compartments.. BELOW: These small hatches allow the operators to easily access specific points of the MCLC after it has been loaded onto the vehicle.

ABOVE: The rear view of the MCLC compartment looking into the compartment. BELOW: The interior of the MCLC compartment, with one inert training charge (blue box) ready to fire.

ABOVE: The top of the rear of the ABV, showing the loaded LMS system and variou sstorage boxes. BELOW: The bottom of the MCLC area, showing the jettison system. This allows the tank commander to "eject" or roll the MCLC from the turret. This feature is only utilized in emergency situations and acts as a last resort.

ABOVE: The camera (in the rectangular box in the center of the MCLC) seen here is one of many that are placed around the vehicle which the operators can choose to monitor. This definitely comes in handy when trying to back up. BELOW: Another look at just how the MCLC is secured to the turret with the four clamps.

ABOVE: The turret rotated 90 degrees showing the LMS system attached to the vehicle. BELOW: A close-up of the right side LMS. Note the hydraulic system allows the tank commander to deploy the LMS from inside the turret. It can be manually deployed as well.

PAGE 45 ABOVE and BELOW: Two close-up views of the LMS system.

ABOVE and BELOW: Two views of the rear of the ABV. The ABV uses the M1 chassis as the basis for the lower hull.

ABOVE: Trainees get ready to conduct their daily inspections before heading out for another day of mastering their skills on this beast. BELOW: The #7 skirt is missing showing the drive sprocket. Note the dirt build-up on the sponson and on the tracks. The track-pads on this vehicle have just about been worn down to the metal track.

Page 48 - 49: Various photos showing the ABV's running gear and the accumulation of dirt and muck during operations in the field. Notice how the wheels and other areas of the lower hull get filthy, while the side skirts get little to no dirt and mud on them.

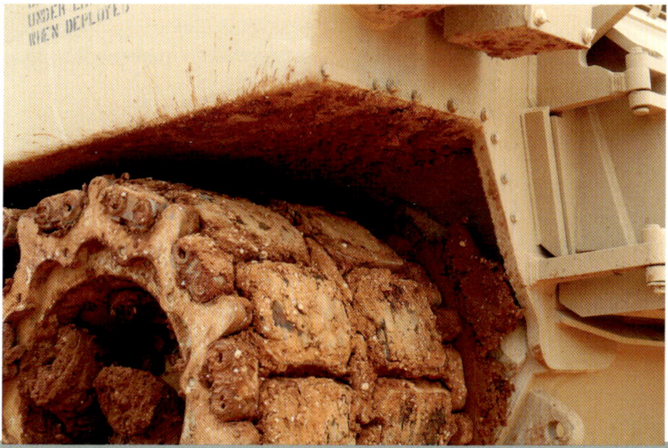

ABOVE: The commander's hatch with a CVC helmet connected to the communications unit. BELOW: Crew of the M1 ABV at Ft. Leonard Wood. Note the driver, PVT Brittany Coffey, who is a little shorter than most crewmen. She used a child's jump seat in the driver's compartment so she could reach the pedals and other instruments. Don't let this fool you, though. She was one tough cookie!

ABOVE: The cradle on the tank commander's hatch allows the ABV to have an M2 .50 caliber machine gun mounted, which the tank commander may fire from inside of the turret. BELOW: Extra tracks are a must in order for operators to perform emergency track maintenance on the fly.

ABOVE: Either side of the turret is equipped with a smoke grenade launcher which can provide screening for other vehicles and dismounts during breaching operations. Note, these are the Army versions.

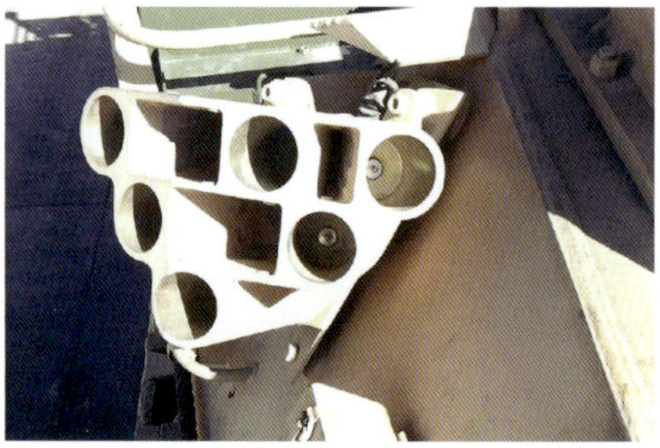

ABOVE: Standard Army smoke grenade lanuncher found on the ABV and other M1-series tanks.

LEFT: USMC grenade launcher configuration found on most USMC vehicles, including Marine ABV's and M1-series tanks.

PAGE 57: Two close-up views of the FWMP atatchment to the front hull of the ABV. The mine plow is getting ready to be attached to the ABV.

B510

B510

TDC 1BDE
35E
B508

B512
LIFT
HERE
AABV009

A Staff NCO gives instruction to the soldiers as they fire the M2 .50 caliber machine gun from the ABV.

Page 60: A nice rear view of the ABV.

RIGHT: Gunnery practice using the M2 .50 caliber machinegun. The M2 can be operated both outside and inside the vehicle.

LASER SAFETY CODE B
INDIVIDUAL FILTER CODE
SERIAL NUMBER

SQUIRE
U.S. ARMY

PAGE 62 ABOVE: With the Common Remotely Operated Weapon Station (CROWS) attached to the M2, the Army's newest spotter shows just how easily the weapon system can be employed.

PAGE 62 BELOW: Soldiers at Fort Leonard Woods's Heavy Track Operator's Course strike a pose to show off their newly acquired ABV.

ABOVE: An ABV from A Company, 588th Brigade Engineer Battalion, 3rd Armored Brigade Combat Team, 4th Infantry Division participating in Operation Atlantic Resolve, Poland, January 2017.

RIGHT: What do soldiers have to say about the ABV? Two thumbs up!